Cambridge Elements

Elements in Geochemical Tracers in Earth System Science

MAGNESIUM ISOTOPES

Tracer for the Global Biogeochemical Cycle of Magnesium Past and Present or Archive of Alteration?

Edward T. Tipper
University of Cambridge

CAMBRIDGE
UNIVERSITY PRESS

University Printing House, Cambridge CB2 8BS, United Kingdom

One Liberty Plaza, 20th Floor, New York, NY 10006, USA

477 Williamstown Road, Port Melbourne, VIC 3207, Australia

314–321, 3rd Floor, Plot 3, Splendor Forum, Jasola District Centre, New Delhi – 110025, India

103 Penang Road, #05–06/07, Visioncrest Commercial, Singapore 238467

Cambridge University Press is part of the University of Cambridge.

It furthers the University's mission by disseminating knowledge in the pursuit of education, learning, and research at the highest international levels of excellence.

www.cambridge.org
Information on this title: www.cambridge.org/9781108994309
DOI: 10.1017/9781108991698

First published 2022

A catalogue record for this publication is available from the British Library.

ISBN 978-1-108-99430-9 Paperback
ISSN 2515-7027 (online)
ISSN 2515-6454 (print)

Magnesium Isotopes

Tracer for the Global Biogeochemical Cycle of Magnesium Past and Present or Archive of Alteration?

Elements in Geochemical Tracers in Earth System Science

DOI: 10.1017/9781108991698
First published online: February 2022

Edward T. Tipper
University of Cambridge
Author for correspondence: Edward T. Tipper, ett20@cam.ac.uk

Abstract: Magnesium is a major constituent in silicate and carbonate minerals, the hydrosphere and the biosphere. Magnesium is constantly cycled between these reservoirs. Since each of the major planetary reservoirs of magnesium has different magnesium isotope ratios, there is scope to use magnesium isotope ratios to trace (1) the processes that cycle magnesium at spatial scales from the entire planet to microscopic and (2) the relative fluxes between these reservoirs. This Element summarises some of the key motivations, successes and challenges facing the use of magnesium isotopes to construct a budget of seawater magnesium, present and past.

Keywords: Mg isotopes, oceanic budget, seawater chemistry, rivers, hydrothermal circulation

ISBNs: 9781108994309 (PB), 9781108991698 (OC)
ISSNs: 2515-7027 (online), 2515-6454 (print)

Contents

1 Introduction

Magnesium (Mg) is the fourth most abundant species in seawater [1], and the eighth most abundant element in the continental crust [2]; it is a major constituent of both silicate and carbonate minerals. Its transfer from rocks to the hydrosphere via chemical weathering on the continents and its return to rocks via (1) exchange with hydrothermal fluids at mid-ocean ridges [3, 4], (2) through carbonate sedimentation [5] and (3) through the formation of clays in the ocean (sometimes referred to as reverse weathering) [6] constitutes one of the major planetary global biogeochemical cycles. Secular variations in the Mg concentration of seawater over geological time, inferred from either fluid inclusions or the Mg/Ca ratio of marine carbonates, likely record changes in the balance between the inputs and outputs of Mg, providing an archive of this major geochemical cycle through Earth history. These records of oceanic Mg/Ca through time have been significantly refined in the last decade [7–15]. This is of fundamental significance to planetary history. For example, the rate of mid-ocean ridge spreading determines the rate of removal of Mg from seawater via modern hydrothermal circulation, one of the key mechanisms for cooling the oceanic crust [16]. The Mg/Ca ratio of seawater has a key role in regulating the main carbonate mineralogy from the ocean, aragonite, calcite or dolomite [5, 17], and as such it exerts a major influence on the biosphere. The chemical weathering of Mg-bearing silicates on the continents and the removal of Mg from seawater via low-temperature hydrothermal circulation may play a fundamental role in regulating planetary temperatures [18, 19], providing a negative temperature-dependent climate feedback.

Despite the fundamental role of oceanic Mg in planetary evolutionary history, the relative importance of the main inputs and outputs of Mg from the oceans remains debated, controversial and difficult to precisely constrain. For example, the hydrothermal sink of Mg has been estimated to range from 40 to 100 per cent of the total flux of Mg from the oceans relative to the dolomite sink that could vary from 0 to 60 per cent [17, 20, 21].

A potentially transformative tool to resolve the oceanic budget of Mg, both today and through planetary history, is the measurement of Mg stable isotope ratios. Mg has three stable isotopes, ^{24}Mg, ^{25}Mg and ^{26}Mg, in the relative abundances of 78.99, 10 and 11.01 per cent respectively [22]. Although first measured in the 1960s [23], routine, reliable and precise measurements only became common in the year 2000 with the development of new multi-collector, inductively coupled plasma mass spectrometers (MC-ICP-MS) [24]. The use of Mg isotopes to constrain the oceanic budget of Mg was first attempted in 2006, presenting initial data of the inputs and outputs and summarising the problem [25]. Since that time, thousands of new measurements have been published, including records through time of marine carbonates thought to trace past seawater, with the potential to unlock the decades-old controversy of the global biochemical cycle of Mg. However, the records are still discrepant, and the extent to which diagenesis could influence some records remains underexplored [26]. The purpose of this Element is to provide a progress update on the oceanic budget of Mg, both in the present day and through the Cenozoic era, from a Mg isotope perspective. The pitfalls in the data and analysis are discussed, but the potential of Mg isotopes, a proxy still in its infancy, as a tracer to decipher seawater history and as one of the most important global biogeochemical cycles through time is emphasised.

Box 1 Mg Isotope Notation

Deviations in the $^{26}Mg/^{24}Mg$ and $^{25}Mg/^{24}Mg$ ratios are expressed in per mil (parts per thousand) notation as:

$$\delta^{X}Mg = 1000 \cdot \left\{ \frac{\left(\frac{^{X}Mg}{^{24}Mg}\right)_{sample}}{\left(\frac{^{X}Mg}{^{24}Mg}\right)_{standard}} - 1 \right\}, \tag{1}$$

where X refers to to either ^{25}Mg or ^{26}Mg, and the standard used in almost all publications is the Dead Sea Mg (DSM3) standard.

A prerequisite of analysis is that only solutions of pure Mg can be introduced into the mass spectrometer (< 5% other contaminant ions). Pure solutions are typically obtained by ion chromatography through many different methods. One difficulty is that Mg isotope ratios are fractionated by the chromatographic procedure, and that no Mg can be lost. It is therefore essential to continually process standards through the entire chromatographic process [27, 28]. Many studies collect a pre- and post-cut on either side of the Mg elution from column chemistry to verify that no Mg is lost. The majority of studies report the results of such standards with a similar matrix (chemical composition) to the samples. Unlike many other stable isotope isotope systems, the vast majority of analyses have not used a double-spike method (though it is technically possible) [29] because there are only three isotopes available. This makes 100% recovery even more important.

2 What Do Mg Isotopes Promise?

The key promise of Mg isotopes stems from the total terrestrial range in excess of 7‰ (Fig. 1) in the $^{26}Mg/^{24}Mg$ ratio, meaning that on the one hand $\delta^{26}Mg$

values are a valuable tracer of different sources of Mg but on the other hand a tracer of the processes that give rise to the 7‰ terrestrial variation. The average terrestrial silicate composition at the Earth's surface is well represented by the mean value of measurements of silicate rocks at −0.32‰ with well over 1,000 measurements [30] (standard deviation provided in Table 3). The rest of the terrestrial variability has likely developed because of mass-dependent fractionation (Box 2) induced by chemical and/or mineralogical reactions.

Box 2 Mass-Dependent Fractionation

Mass-dependent fractionation is either an equilibrium (quantum mechanical) or a kinetic (classical mechanics) phenomenon that induces preferential transport or incorporation of one isotope relative to another during a chemical or mineralogical reaction. All substances are stabilised by the incorporation of the colloquially named 'heavy isotope', [31] ^{26}Mg in the case of Mg, which lowers the energy of a phase in proportion to the inverse square root of the mass. Some phases, however, can minimise their energy more relative to others because of their bonding environment. A fractionation factor between two phases A and B is defined as the isotopic ratio in phase A divided by the isotopic ratio of phase B when these phases are in chemical equilibrium. This ratio is commonly denoted by the Greek letter α, and the difference in delta notation is approximately equal to:

$$\delta^{26}Mg_A - \delta^{26}Mg_B = 1000 \cdot \ln(\alpha) = 1000 \cdot \ln\left\{\frac{\left(\frac{^{26}Mg}{^{24}Mg}\right)_A}{\left(\frac{^{26}Mg}{^{24}Mg}\right)_B}\right\}. \quad (2)$$

Any fractionation of the $^{25}Mg/^{24}Mg$ ratio is related to the $^{26}Mg/^{24}Mg$ ratio but is scaled by 0.520 (in the case of equilibrium fractionation) and known as mass-dependent fractionation [32]. There have been major advances in (1) the quantum mechanical calculation of equilibrium fractionation factors in the past two decades [33, 34] (known as density functional theory) and (2) experimental studies that determine fractionation factors [35, 36]. Key findings are that all carbonate minerals and some clays have an affinity for ^{24}Mg [34, 36], meaning that carbonate minerals have a lower $\delta^{26}Mg$ than the aqueous solution they precipitated from. This observation is qualitatively consistent with the distinct Mg bonding environment in the carbonate mineral compared to the fluid phase, as reflected by the average Mg-O distance, with the fluid having shorter (higher vibrational frequency) Mg-O bond lengths that can be stabilised to a greater extent by incorporation of

^{26}Mg. For example, the enrichment of brucite, magnesite and dolomite in ^{24}Mg is consistent with the longer average Mg-O distance in these minerals compared to aqueous Mg^{2+} [37].

Carbonate minerals show the greatest difference or degree of fractionation from silicate rocks. Modern marine carbonates show a very wide range of 6‰ for δ^{26}Mg values, depending on mineralogy, such as dolomite (−1.9‰), high-Mg calcite (−2.6‰), low-Mg calcite (−3.7‰) and aragonite (−1.9‰). These contemporary marine carbonate phases have formed from modern seawater which has a very tightly defined value of −0.82‰, demonstrating the affinity of ^{24}Mg for all carbonate minerals (see Box 2 for an explanation and Box 3 for a more detailed summary). The differences in the δ^{26}Mg of carbonate minerals almost certainly arise from (1) differences in the bonding environments of Mg within the different carbonate minerals, (2) the mechanism of incorporation of Mg into the carbonate mineral (biomineralisation). For example, the calcitic microorganisms foraminifera and coccoliths both construct their tests from low-Mg calcite, but their δ^{26}Mg typically differs by > 2‰ (Box 3) [38]. This difference is likely caused by biological processes during bio-crystallisation and illustrates the potential of Mg isotopes to trace and understand those processes. Thirdly, crystallisation rate (kinetics) has been shown to exert a major influence on the δ^{26}Mg of inorganic calcite [35]. A higher precipitation rate reduces the difference between δ^{26}Mg in the carbonate mineral and the fluid. This is thought to be because the limiting step for carbonate precipitation is the dehydration kinetics of the Ca and Mg ions. These dehydration kinetics are thought to be five orders of magnitude slower for Mg than for Ca, meaning that very slow precipitation rates are required to reach isotopic equilibrium in principle [35]. It is hoped that Mg isotopes will provide a window into these precipitation mechanisms.

The terrestrial hydrosphere (defined here as river waters, groundwaters, soil pore waters and cave drip waters) also shows a wide range in δ^{26}Mg values from approximately 0 to −3‰ with a mean δ^{26}Mg value of −1.08‰ (Fig. 1, Box 3) [25]. This composition of terrestrial waters is intermediate between carbonate and silicate minerals (the principal sources of aqueous Mg through mineral dissolution) and highlights one of the key potential applications of Mg isotopes: quantifying the proportion of carbonate and silicate Mg released to the hydrosphere. This variability in sources is present only because of mass-dependent fractionation in the first place, but this fractionation is also a challenge for partitioning sources. In addition to the source effect arising from carbonate and silicate rocks, δ^{26}Mg values are subject to mass-dependent fractionation. For example, both the uptake of Mg in the biosphere by plants [39] and the formation of secondary mineral phases on the continents during chemical weathering fractionate Mg isotopes, meaning that Mg isotopes might make a useful tracer of these processes. Many aspects of geochemistry are challenged by the source-versus-process issue, and the Mg isotope budget of the modern ocean is an ideal illustration of this problem.

Box 3 Summary of the δ^{26}Mg Values of the Main Terrestrial Reservoirs

Type	Mineralogy	median	mean	Standard deviation	Max	Min	n
Marine Carbonate		−3	−3.3	1.2	−1	−5.6	551
Sponge	Aragonite	−2.6	−2.4	0.7	−3.2	−1.5	13
Coral	Aragonite	−1.8	−1.8	0.1	−2.2	−1.5	80
Coral	HMC	−3.5	−3.5	0.1	−3.5	−3.4	2
Foraminifera	HMC	−3.3	−3.1	0.4	−3.7	−2.5	9
Sponge	HMC	−3.2	−3.2	0.1	−3.4	−3.2	6
Coralline algae	HMC	−3.2	−3.2	0.1	−3.2	−3	9
Echinoderm	HMC	−2.2	−2.1	0.6	−2.8	−1.3	30
Foraminifera	LMC	−4.7	−4.7	0.4	−5.6	−4	83
Mollusc	LMC	−4.6	−4.4	0.5	−5.1	−3.4	15
Echinoderm	LMC	−3	−3.2	1	−5.5	−1.3	55
Brachiopod	LMC	−2.3	−2.3	0.5	−4	−1.4	57
Coccolithophore	LMC	−1.9	−1.9	0.9	−3.1	−1	8
Dolomite		−2	−1.9	0.6	0.4	−3.8	412
Terrestrial Water		−1	−1.1	0.5	0.6	−3.9	831
Seawater		−0.8	−0.8	< 0.1	−0.6	−0.9	71
Silicate Rock		−0.3	−0.3	0.4	1	−3.1	1206
Soil		−0.2	−0.1	0.5	1.8	−1.3	252

Note: LMC refers to low-Mg calcite, and HMC refers to high-Mg calcite.

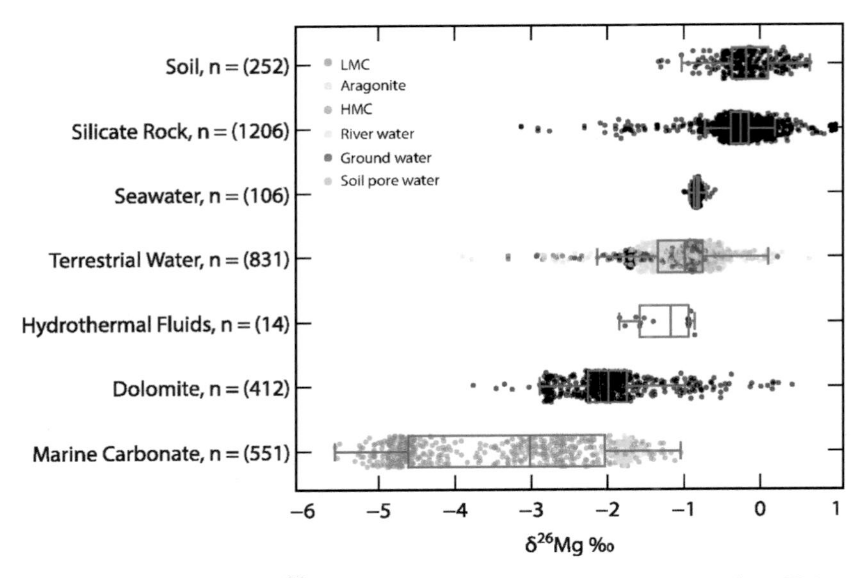

Figure 1 Summary of δ^{26}Mg values in the main terrestrial reservoirs of Mg.

Box 4 Fractionation Associated with Clay Minerals (Phyllosilicates)

The direction and magnitude of magnesium (Mg) isotope fractionation associated with the formation of clay minerals is fundamental to the use of Mg isotopes to decipher the biogeochemical cycling of Mg both in the critical zone and for the oceanic Mg budget. Clay minerals, which form in both the continental and marine realms, contain appreciable amounts of Mg, and several recent studies have highlighted the formation of marine clays as being key regulators of both the global cycle of Mg and the carbon cycle through reverse weathering processes [6, 40]. Some of the first studies on riverine Mg noted that river waters draining first-order catchments with only silicate rocks are nearly always enriched in ^{24}Mg relative to silicate rocks [41, 42]. This was interpreted to result from the formation of clay minerals preferentially incorporating ^{26}Mg. This is consistent with soil showing a subtle enrichment in ^{26}Mg in soils at a global scale (Fig. 1) [39, 43, 44]. However, in contrast, at a local scale some field studies have inferred that clays are enriched in ^{24}Mg [45–47].

There are at least two potential mineralogical sites for Mg in a clay mineral: (1) structurally bound Mg in octahedrally coordinated lattice sites and (2) loosely bound adsorbed or exchangeable Mg sites associated with surface or interlayer negative charges. Each of these sites may have its own fractionation factor. Some field studies have endeavoured to distinguish fractionation factors associated with each mineralogical site [48, 49]. Experimental work has also tried to determine the fractionation factors associated with clays and to distinguish between different lattice sites. In addition, phyllosilicate minerals kerolite and lizardite are enriched in ^{26}Mg [50] whereas hectorite, chrysotile and brucite (with one exception) [51] are enriched in ^{24}Mg [36, 37].

These contrasting enrichments in ^{24}Mg or ^{26}Mg are predicted from the relative bond lengths of these minerals.

3 The Modern Oceanic Budget System

There is a large oceanic reservoir of Mg with a concentration of 53 mmol/l. This means that the residence time (defined as the ratio of the reservoir size to the output flux) is long (> 10 Ma) and much greater than the mixing time of the oceans (~1,000 yrs), meaning that both the concentration and the isotopic composition of the ocean are uniform. The isotopic composition of seawater is perhaps the most homogenous reservoir of Mg on the planet with a δ^{26}Mg of −0.82‰. This concentration and isotopic composition is maintained by inputs (rivers and groundwaters) and outputs (hydrothermal circulation and carbonate formation) as summarised in Fig. 2.

3.1 Inputs to the Modern Ocean

Mg is continually supplied to the oceans by rivers [25] and groundwaters [52], with average δ^{26}Mg of −1.08‰. This is very similar to the flux weighted mean of large rivers where ~50 per cent of the flux was sampled [25]. This agreement between the flux weighted mean and the overall average of terrestrial fresh waters is encouraging, suggesting that the input flux to the oceans is well constrained in the modern system. However, the wide range of terrestrial waters (Fig. 1) means that there is scope for the terrestrial δ^{26}Mg to have changed in the past. It is noteworthy, for example, that rivers draining basalt have δ^{26}Mg values that are higher than the terrestrial mean value [45], giving rise to the possibility that during the emplacement and weathering of large continental flood basalt provinces the δ^{26}Mg of the continental input to seawater may adjust to higher values. Alternatively, during periods of orogenesis, when continental margin sedimentary sequences tend to be exhumed and weathered, a shift to a greater amount of carbonate weathering would lower the δ^{26}Mg of the continental output.

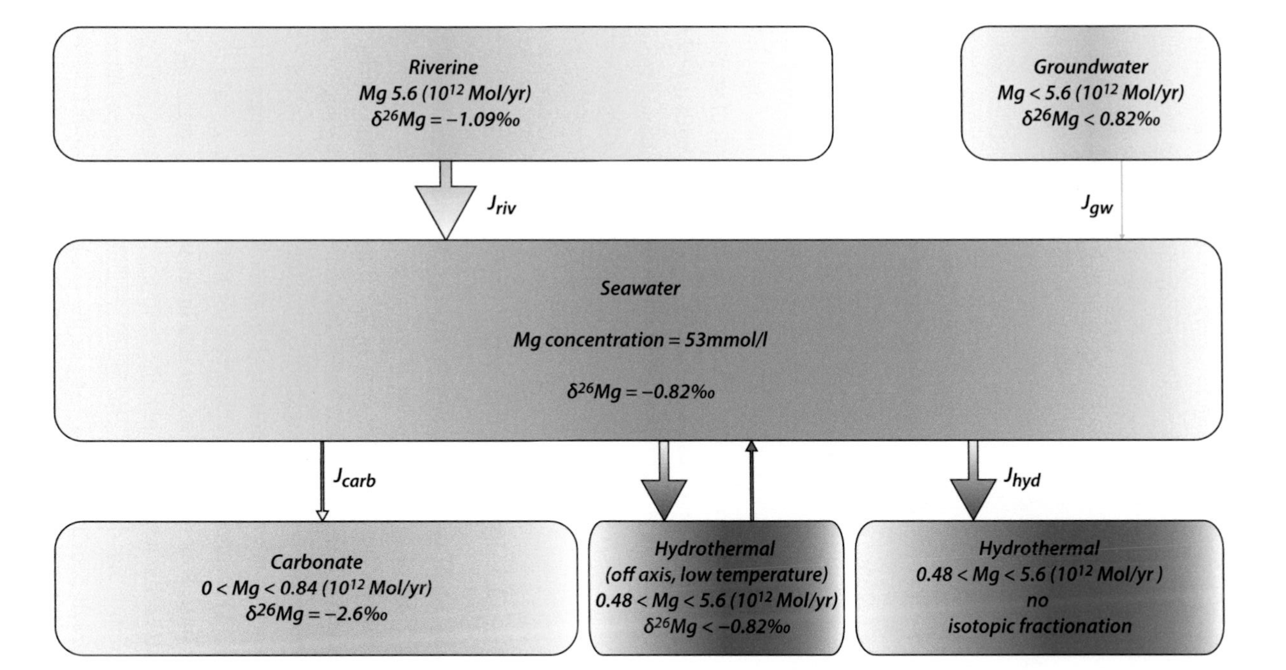

Figure 2 Summary of the modern oceanic budget of Mg showing the main fluxes to and from the modern ocean, indicating their magnitude in Tmol/yr and isotopic composition in per mil.

In the modern system, the δ^{26}Mg value of the continental input to seawater is lower than seawater δ^{26}Mg values by $< 0.3‰$. This relatively small difference can only be explained by one or a combination of three mechanisms:

1. There is an additional supply of Mg enriched in ^{26}Mg.
2. The processes which remove Mg from the ocean fractionate Mg isotope ratios.
3. The current system is not in steady state.

These mechanisms are explored below.

3.2 Outputs

Mg is continually removed from seawater by (1) hydrothermal circulation and (2) carbonate formation (Fig. 2). Hydrothermal circulation can be distinguished into two forms: high-temperature hydrothermal flow close to the ridge axis, and low-temperature, off-axis hydrothermal flow (Fig. 2) [17]. The removal of Mg from seawater via high-temperature hydrothermal circulation is thought to be quantitative because hydrothermal fluids collected from black smokers are highly depleted in Mg (< 5 per cent of the seawater value) [53]. Because this removal of Mg from seawater is quantitative, it is not accompanied by an isotopic fractionation because there is no return flux to seawater. Therefore, Mg removed via high-temperature hydrothermal circulation is removed with the δ^{26}Mg value of seawater itself. The relatively small difference in the δ^{26}Mg values of the inputs to seawater and seawater itself of $< 0.3‰$ is consistent with high-temperature hydrothermal circulation being the dominant sink of Mg, with no isotopic fractionation [17].

Low-temperature, off-axis circulation is, on the other hand, not quantitative. Measurements of δ^{26}Mg in low-temperature hydrothermal fluids from the Juan de Fuca and Cocos plates yield δ^{26}Mg values that are strongly offset to values lower than that of seawater (Fig. 1) [17]. Because of the difficulty in sampling hydrothermal fluids, it was previously hypothesised that the Mg in off-axis hydrothermal fluid samples could have resulted from seawater contamination. However, the significant offset in δ^{26}Mg values between low-temperature hydrothermal fluids and seawater implies that even if there is seawater contamination during sampling, there must be a process that fractionates Mg isotope ratios during low-temperature hydrothermal circulation. Continental hydrothermal fluids flowing through Icelandic basalts are affected by a similar process [54].

This process is likely the formation of low-temperature secondary minerals, in particular smectites (clays), within the oceanic crust. The enrichment of low-temperature hydrothermal fluids in ^{24}Mg is consistent with measurements of altered oceanic crust from the Troodos ophiolite and altered pacific MORB [55], which are enriched in ^{26}Mg. The samples from the Troodos contain up to 66 per cent smectite, perhaps indicating an unusually high degree of alteration from MORB. Some of this chemical/mineralogical alteration could have occurred during post-tectonic exhumation. Nevertheless, this pairing of ^{24}Mg-enriched fluids with ^{26}Mg-enriched altered oceanic crust suggests that in the field, the smectites associated with hydrothermal circulation have an affinity for ^{26}Mg (see Box 4 for further discussion about the Mg isotope fractionation factors into phyllosillicates). The net result of hydrothermal circulation is therefore that the inputs to seawater are more enriched in ^{24}Mg than seawater itself. This fractionated removal of ^{26}Mg to the oceanic crust effectively means that there is a return flux of Mg to seawater that is enriched in ^{24}Mg. This would drive the difference between the $\delta^{26}Mg$ values of the inputs to seawater and seawater itself to $> 0.3‰$.

Carbonate formation provides an additional sink of Mg from seawater (Fig. 1). It is likely that the vast majority of Mg removal via carbonate is via dolomite because of the high Mg/Ca ratio, close to 1 in stoichiometric dolomite. In the modern system, dolomite formation would qualitatively account for the offset in $\delta^{26}Mg$ between the riverine input and seawater of $< 0.3‰$ (greater if the return flux from low-temperature hydrothermal circulation is accounted for). Since the riverine input has a lower $\delta^{26}Mg$ value than seawater, dolomite, enriched in ^{24}Mg on average by $> 1‰$ relative to seawater, could be enough to account for the higher $\delta^{26}Mg$ value of seawater compared to rivers. Tipper et al [25] estimated that the minimum dolomite flux from modern seawater was 8 per cent and as high as 40 per cent. More recently, Shalev et al [17]. argued for an even greater removal of Mg (40–60 per cent) via dolomite after accounting for the low-temperature hydrothermal removal of ^{26}Mg.

In the modern system, dolomite formation has been a long-standing conundrum since the kinetics of dolomite precipitation are slow, and there is a high energy of nucleation [5]. Prior to the mass balance constraint from Mg isotopes, some authors argued that the total proportion of dolomite removal of Mg in the modern system was < 10 per cent. The fraction of Mg removed from the oceans via dolomite relative to hydrothermal circulation has been contentious with regard to the modern ocean, but is thought to be one of the key drivers in changes in the Mg content of the ocean over geological history [5]. It is therefore possible that the reason for the offset in $\delta^{26}Mg$ between the inputs and outputs is because the oceans are not at steady state.

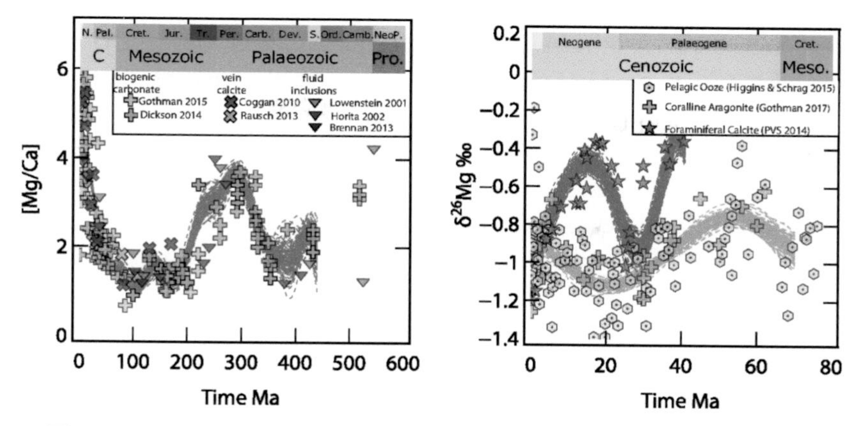

Figure 3 Summary oceanic records of Mg cycling. Left: Changes in the marine Mg/Ca ratio though time determined from biogenic calcite [7, 8], vein calcite [9, 10] and fluid inclusions [11–14]. The grey lines show multiple (Monte Carlo) Fourier series fits through the data, based on the uncertainties in the published measurements and ages. Right: Records of δ^{26}Mg in seawater (note the different timescale) [56–58]. The lines through the foraminifera δ^{26}Mg record (blue) show smoothed Fourier series best fits (third order with a period of 25 Ma) through the data, taking into account the uncertainty in the measurements, the fractionation factor and the age. Fourier series were fitted through the pelagic ooze [57] and coralline aragonite [58] together using a periodicity of ~130 Ma.

4 Past Records of Seawater

The oceanic budget of Mg throughout Earth history is likely dynamic. The Mg/Ca ratio of seawater, as reconstructed from fluid inclusions [11–14], vein calcite [9, 10], and biogenic calcite [7, 8], shows long-wavelength (~100 Ma) secular variations with a high Mg/Ca ratio of ~ 4 during the late Palaeozoic/early Mesozoic, a low throughout most of the Mesozoic, followed by a rapid increase to the present-day ratio of ~5 over the last 100 Myr (Fig. 3). These changes in Mg/Ca ratio are a testament to global changes in the Mg budget of seawater over time. The expectation is that Mg isotope ratios may serve as a proxy to help decipher the causal mechanisms of these major changes in the biogeochemical cycle of Mg. There are to date three Cenozoic records of δ^{26}Mg values in seawater through time, reconstructed from marine carbonates (Fig. 3B).

4.1 Seawater δ^{26}Mg Records from Marine Biogenic Carbonates

4.1.1 Foraminiferal Calcite

The first published seawater record of δ^{26}Mg used measurements of δ^{26}Mg in foraminifera [56] to estimate seawater δ^{26}Mg values through time. Individual foraminifera species *G. tumida, G. ruber, G. sacculifer, O. universa, G. trunclinoides, G. inflata* and *G. conglobatus* were picked and cleaned from core tops from sites in the Southern Atlantic, and from drill cores from the Walvis ridge in the Southern Atlantic. The modern fractionation factor between foraminiferal calcite and seawater was estimated as $\sim -4.6‰$ using recent foraminifera from the core tops and by comparing to the δ^{26}Mg value of modern seawater [38, 59]. It was also demonstrated (with few exceptions) that there were no appreciable differences between species, making it possible to use different species of foraminifera to generate a record through time – an important consideration, since one single species is not available to construct a record throughout the entire Cenozoic. All foraminifera were cleaned to remove both clays and organic material that could contaminate the δ^{26}Mg of the calcite itself. Only specimens with minimal overgrowth or diagenetic infilling were analysed, and trace element ratios such as manganese to calcium and aluminium to calcium were measured to demonstrate the efficacy of the cleaning procedure.

Two species of foraminifera, *G. venezuelana* and *O. universa*, were picked from core samples ranging in age from the Eocene to the Pleistocene and analysed for δ^{26}Mg values. The resulting seawater δ^{26}Mg record shows significant temporal variability of $> 0.8‰$ on a timescale of $\sim$ 10 Ma (Fig. 3). The record suggests that seawater δ^{26}Mg values were generally higher in the past than they are in the present day by approximately 0.4‰, but with a sharp spike towards present δ^{26}Mg values at the end of the Palaeogene (albeit based on only three samples).

4.1.2 Bulk Pelagic Ooze

The second and most extensive published seawater record of δ^{26}Mg values was based on bulk pelagic ooze (comprised of foraminifera and coccoliths), from the Walvis ridge, and the Ontong Java Plateau in the Pacific [57]. Although foraminiferal calcite has a fractionation factor from seawater that is different to that from coccoliths by >2‰ (Box 3), Higgins and Schrag [57] argue that it is possible to extract a seawater record rather than a mixing proportion of foraminifera to coccoliths because foraminifera contain greater than a factor of 10 more Mg relative to Ca content compared to coccoliths. In addition, two size fractions were analysed, < 65 microns, inferred to be predominantly coccoliths, and 250–450 microns, inferred to be foraminifera, and no systematic differences were observed between the size fractions, bringing into question whether there really is a difference in fractionation factor between coccoliths and foraminifera. Further, some bulk limestones were also analysed for the older parts of the record. Pore waters were also analysed for δ^{26}Mg for some samples [57]. A spatially and temporally resolved box model of the pore fluid and pelagic ooze δ^{26}Mg values was used to demonstrate that diagenesis has a minimal impact on δ^{26}Mg (discussed in greater detail in Section 5). The modelling shows that the bulk sample could have shifted to marginally lower ($<$ 0.3‰) or higher ($<$0.1) δ^{26}Mg values, depending on the extent and precise parameters of diagenesis. This record, which extends back to the late Mesozoic, is markedly different to the foraminiferal calcite record (Fig. 3B). Firstly, the record suggests that seawater had a lower δ^{26}Mg value compared to the present. Secondly, there are no high-frequency variations, but rather longer wavelength periodicity to the data ($\sim$130 Ma), with a slight decrease in the early Cenozoic, with the estimated δ^{26}Mg of seawater decreasing to values approximately 0.2‰ lower than present values and then gradually increasing over the last 20 Ma.

4.1.3 Coralline Aragonite

The third published seawater record of δ^{26}Mg was based on a pan-global collection of un-recrystallised aragonitic fossil scleractinian corals [8, 58], carefully screened for diagenesis using X-ray diffractometry, cathodoluminescence microscopy, and Raman spectroscopy to demonstrate little evidence of secondary calcite. In addition, strontium isotope ratios, clumped isotopes, and trace element ratios sensitive to diagenesis such as the manganese to calcium ratio were also measured to test for geochemical changes in diagenesis. The aragonite δ^{26}Mg values are converted to seawater δ^{26}Mg assuming a constant fractionation factor for coralline aragonite through time of −0.9‰ [38, 58, 60]. The coralline aragonite seawater δ^{26}Mg record is remarkably consistent with the bulk pelagic ooze record (the fitted lines on Fig. 3B are through both data sets).

4.1.4 Interpretation of Marine Records

It is not clear why the foraminiferal δ^{26}Mg seawater record is so different to the bulk pelagic ooze and coralline aragonite records, which in turn are similar. It is equally not clear at this point which, if any, is the most reliable record. If either record is correct, a consistent interpretation of seawater δ^{26}Mg values and Mg/Ca ratios over time will be different. Whilst it is possible to attempt to solve the system for a unique solution, here the spectrum of possible interpretations is illustrated for the oceanic budget of Mg as a sensitivity study through time. This illustrates the potential utility of the Mg isotope proxy rather than providing a solution.

As discussed for the modern oceanic budget, the past oceanic budget can be expressed as a simple box model though time (Fig. 2). In this simplistic case, the change in seawater Mg concentration and δ^{26}Mg values through time can be expressed as first-order differential equations:

$$\frac{dN_{Mg}}{dt} = J_{riv} - J_{hyd} - J_{dol} \tag{3}$$

$$N_{Mg} \cdot \frac{d\delta_{sw}}{dt} = \delta_{riv} J_{riv} - \delta_{hyd} J_{hyd} - \delta_{dol} J_{dol}. \tag{4}$$

Equation (3) describes the rate of change in mass of oceanic Mg (N_{Mg}) over time (t) as a function of the difference between the riverine input (J_{riv}) and output fluxes via dolomite and hydrothermal circulation (J_{dol} and J_{hyd}). Equation (4) describes the change in Mg isotopic composition of seawater (δ_{sw}) over time as a function of the input and output fluxes, but also their isotopic compositions (riverine δ^{26}Mg δ_{riv}, hydrothermal δ^{26}Mg δ_{hyd}, and dolomite δ^{26}Mg δ_{dol}). Equations (3) and (4) describe the system in the most simplistic manner possible. There is, for example, no feedback between the flux of Mg leaving the ocean and the concentration of Mg in the ocean. In the natural system, the flux of Mg leaving the ocean likely increases with increasing oceanic Mg content, acting to stabilise changes in the Mg content through time. As written, Eq. (4) assumes that all hydrothermal circulation is high temperature and that removal of Mg is not accompanied by an isotopic fractionation. Both these simplifying assumptions could be accounted for by modifying Eq. (4). An additional simplifying assumption is that the total reservoir of oceanic Mg, N_{Mg}, is proportional to ^{24}Mg.

Assuming that the Ca concentration of the ocean stayed constant though time (an assumption which may or may not be correct in the first order) [15, 61], the increasing Mg/Ca ratio over the past 60 Ma implies that the system was not at steady state. The input flux of Mg must have been greater than the output flux of Mg. Therefore time-dependent solutions to Eqs. (3) and (4) are required. This budget of a non-steady state is a function of the flux imbalance between the inputs and outputs of oceanic Mg, their isotopic composition, and the relative balance of the two main sinks of Mg from the ocean which exert a leverage on the oceanic budget of Mg.

Solutions to Eq. (3) for different degrees of imbalance between the input and output fluxes ($\frac{J_{riv}}{J_{hyd}+J_{dol}} = J_{in}^{Mg}/J_{out}^{Mg}$, contour lines) are indicated in Fig. 4A, enveloping the data of the Mg/Ca ratio of seawater during the Cenozoic. These contour lines of changing Mg/Ca schematically illustrate the behaviour of the system, with, for example, the data enveloped between J_{in}^{Mg}/J_{out}^{Mg} values of 1 and 1.2. However, if, for example, the increase in Mg/Ca starts later at 30

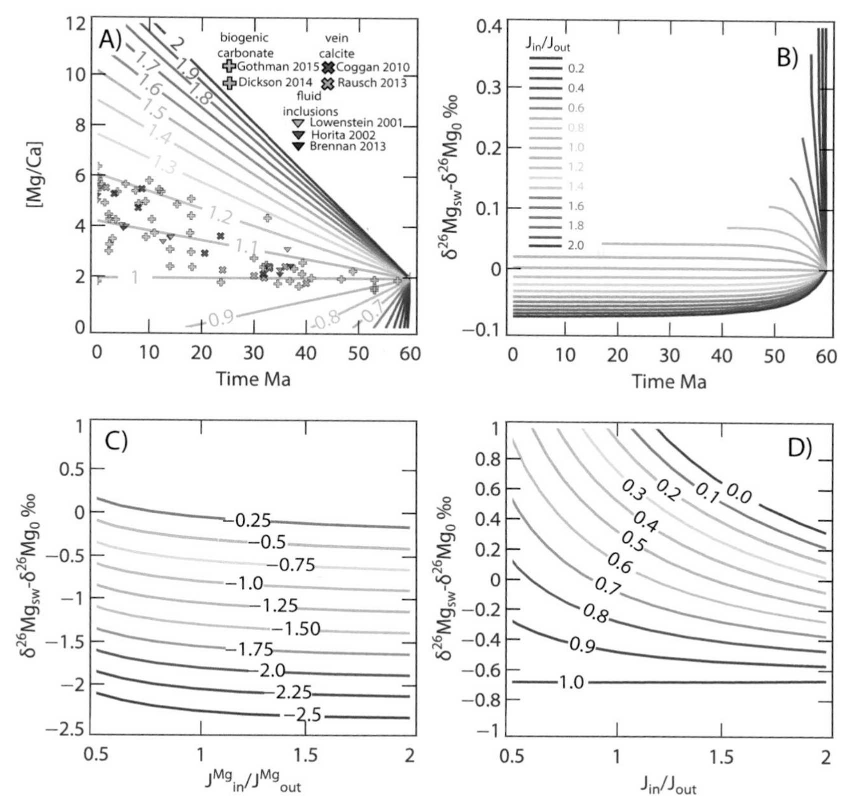

Figure 4 Box-model results of the sensitivity of the oceanic Mg budget. (A) Changes in the Mg/Ca ratio of seawater over time, predicted from Eq. (3) and contoured for J^{Mg}_{in}/J^{Mg}_{out}. The marine proxy data from Fig. 3A is also shown. (B) Changes in the δ^{26}Mg value of seawater (expressed relative to modern seawater, $\delta^{26}Mg_0$), contoured for flux imbalance J^{Mg}_{in}/J^{Mg}_{out}. (C) The sensitivity in the δ^{26}Mg value of seawater to the δ^{26}Mg of the riverine input (contour lines), as a function of the imbalance between the input and output fluxes J^{Mg}_{in}/J^{Mg}_{out}, whilst keeping $\frac{J_{dol}}{J_{hyd}+J_{dol}} = 0.1$, a plausible value for the modern system. (D) The sensitivity in the δ^{26}Mg value of seawater to the fraction of Mg removed via the dolomite to hydrothermal sink $\frac{J_{dol}}{J_{hyd}+J_{dol}}$ (contour lines), plotted for a δ^{26}Mg value of the riverine input of -1.1‰, close to the value for the modern system.

Ma, the contour lines would need to be steeper to envelop the data (equivalent to there being a greater flux imbalance between the input and output fluxes).

This simple imbalance between the input and output fluxes, changing the concentration of Mg in seawater, is sufficient to impart a change in the δ^{26}Mg value of seawater (Fig. 4B) because each of the input and output fluxes has a different isotopic composition. For example, if the ratio of the dolomite and hydrothermal output fluxes is fixed at a plausible value for the modern system ($\frac{J_{dol}}{J_{hyd}+J_{dol}} = 0.1$), and the riverine δ^{26}Mg is fixed at the modern value (~ -1.1‰), then the change in the δ^{26}Mg value of seawater depends on whether the concentration of Mg in seawater is increasing or decreasing. In the case where seawater Mg content is increasing, the δ^{26}Mg of seawater effectively adjusts closer to the δ^{26}Mg of the riverine input (-1.1‰ in this case), lowering the δ^{26}Mg of seawater relative to the present-day value. In this scenario the only scope for variation in the δ^{26}Mg value of seawater is between the present δ^{26}Mg value and that of the riverine input. In the case where seawater Mg is decreasing (not thought to be the case in the Cenozoic), there is much more scope for variation in the δ^{26}Mg value of seawater, because as the reservoir of seawater Mg becomes smaller, there is effectively an isotopic distillation as carbonate precipitation fractionates δ^{26}Mg values. The distillation ceases in this simplistic box model when Mg is completely depleted from seawater (the point at which the lines stop on Fig. 4B). In the natural system, dolomite would cease to precipitate when the concentration drops below a threshold value as the solution drops below the saturation state of dolomite in pore-water environments where the dolomite is precipitating. This would act as a negative feedback on oceanic Mg concentrations, preventing the Mg content of seawater from being completely depleted.

However, there is no a priori reason that the balance of the output fluxes via hydrothermal circulation and dolomite formation will have remained constant over time, nor that the isotopic composition of the riverine input will have remained constant over time. The sensitivity of seawater δ^{26}Mg to these variables can be addressed using Eq. (4) (Fig. 4C and D). The δ^{26}Mg of the riverine input flux at a global scale may change because of changing the proportion of silicate to carbonate weathering. The response of the oceanic δ^{26}Mg value is intuitive, adjusting to close to the value of the input (shown as the contour lines in Fig. 4C), when the input flux is greater than the output flux (increasing oceanic Mg, with the seawater δ^{26}Mg reservoir effectively being replaced with riverine Mg). When the input flux is low relative to the output flux (decreasing oceanic Mg), the effect of carbonate precipitation becomes important, and the δ^{26}Mg values of seawater are predicted to evolve to higher values.

The relative proportion of the dolomite to the hydrothermal sink (contoured as $\frac{J_{dol}}{J_{hyd}+J_{dol}}$ in Fig. 4D) also has a major impact on the δ^{26}Mg value of seawater over time, since dolomite preferentially incorporates ^{24}Mg, and high-temperature hydrothermal circulation removes Mg with the δ^{26}Mg of seawater. When $\frac{J_{dol}}{J_{hyd}+J_{dol}}$ is high, then the δ^{26}Mg of seawater is much higher because of the preferential removal of ^{24}Mg via dolomite, enriching seawater in ^{26}Mg.

The above sensitivity tests are useful to help qualitatively consider how the marine δ^{26}Mg records over time can be used to interpret why the Mg budget has changed through time, and several key points are apparent:

1. A significantly lower δ^{26}Mg (> 0.3‰) than the present value can only be achieved by lower δ^{26}Mg values of the input to seawater.
2. Lower seawater δ^{26}Mg values than present, as low as the current riverine δ^{26}Mg input, can be achieved by reducing the dolomite sink relative to the hydrothermal sink.
3. Seawater δ^{26}Mg values higher than the present day require either a higher riverine δ^{26}Mg input or greater dolomite formation.

Therefore, it is clear that the contrasting foraminiferal δ^{26}Mg record of seawater compared to the bulk pelagic ooze and coral record of seawater δ^{26}Mg have very different interpretations. Whilst both records must be interpreted in the context of the increasing Mg/Ca ratio of seawater, the foraminiferal record tends to be more consistent with either the riverine δ^{26}Mg value being higher in the past and/or dolomite formation being higher in the past and decreasing throughout the Cenozoic to the present day. In contrast, the pelagic ooze and coral records of seawater δ^{26}Mg tend to suggest a relatively static configuration of the balance of the fluxes in and out of the ocean. The pelagic ooze and coral records are consistent with a decrease in the fraction of removal of Mg via dolomite, or a decrease in the δ^{26}Mg of the riverine input between 60 and 20 Ma, followed by an increase in the fraction of removal of Mg via dolomite, or an increase in the δ^{26}Mg of the riverine input between 60 and 20 Ma. The discrepancy between the records does raise one significant question: do these carbonate proxies record seawater at all?

5 Diagenesis: Not Records of Seawater at All?

The fidelity of any marine proxy record depends on the degree of diagenetic alteration of the host material and on the extent to which a specific proxy system is susceptible to change by recrystallisation. There are now several studies that have explicitly considered the impact of diagenesis on δ^{26}Mg values [62–67]. Diagenesis, defined here as the recrystallisation of a metastable, biogenic, high-Mg carbonate or aragonite to a more thermodynamically stable low-Mg calcite or dolomite, is driven by the ingress of an external fluid. It is influenced by multiple variables, such as the dissolution and recrystallisation rate, the rate of advection of an external fluid relative to the rate of dissolution and recrystallisation (which determines the degree of "openness" of the system), the composition of the external fluid (chemistry and isotopic composition), and the chemistry of the secondary phase that is precipitating, determined by a partition coefficient and fractionation factor (which in turn are rate dependent) [35].

The extent to which diagenesis has influenced Mg isotope proxy records has so far been assessed either by screening of the samples for recrystallisation using electron microscopy and elemental ratios such as manganese:calcium [58], which provide a signature of diagenesis, or by modelling to evaluate the potential extent and direction of change of δ^{26}Mg values [26, 62–64]. Modelling studies have conceptualised diagenesis as a two-box model (which can be extended into many boxes to include a spatial dimension) [62]. In such a model, a fixed starting volume of biogenic limestone has a given porosity filled with pore fluid containing a reservoir of Ca and Mg (N_{Ca} and N_{Mg}). Fluid is advected into and out of the pore fluid reservoir at a given flux, which drives dissolution of the host biogenic limestone at a given rate, and reprecipitation of a secondary phase (Fig. 5). The pore-water solution is modelled to remain saturated with respect to calcite and dissolves and precipitates calcite, maintaining the Ca concentration in the pore-water constant, such that the flux of Ca from dissolution is equal to that from precipitation (Fig. 5). The fluid driving diagenesis could be either seawater or meteoric water. In contrast, the supply of Mg through dissolution is not equal to that from precipitation because the Mg/Ca ratio of the host limestone is different from that of the secondary calcite that is precipitating, because the partition coefficient (that relates the Mg/Ca ratio of the precipitated secondary mineral to that of the fluid) is low [63], such that secondary calcite typically precipitates with a lower Mg/Ca ratio than the biogenic limestone. The leverage of the recrystallisation process to modify the δ^{26}Mg of the limestone therefore relates to the supply of Mg from the pore fluid with a different δ^{26}Mg value to the limestone and the relative amount of Mg incorporated into secondary calcite and its fractionation factor.

Such a box-model conceptualisation of diagenesis is very similar to the oceanic budget of Mg described in Section 5.4.1 and similarly can be approximated by a first-order differential equation:

$$\frac{d\delta_{pf}}{dt} = \frac{1}{N_{pf}}\{J_{adv,in}\cdot\delta_{adv,in} - J_{adv,out}\cdot\delta_{adv,out}\} + \frac{1}{N_{pf}}\left\{J_{diss,in}\cdot\delta_{diss,in} - J_{precip,out}\cdot\delta_{precip,out}\right\}, \quad (5)$$

where N_{pf} is the mass of Mg in the pore fluid reservoir, $J_{adv,in}$ and $J_{adv,out}$ describe the advective flux of Mg in and out of the system, $J_{diss,in}$ and $J_{precip,out}$ describe the Mg fluxes into and out of the pore fluid from dissolution and precipitation, and δ describes the δ^{26}Mg of each of the fluxes.

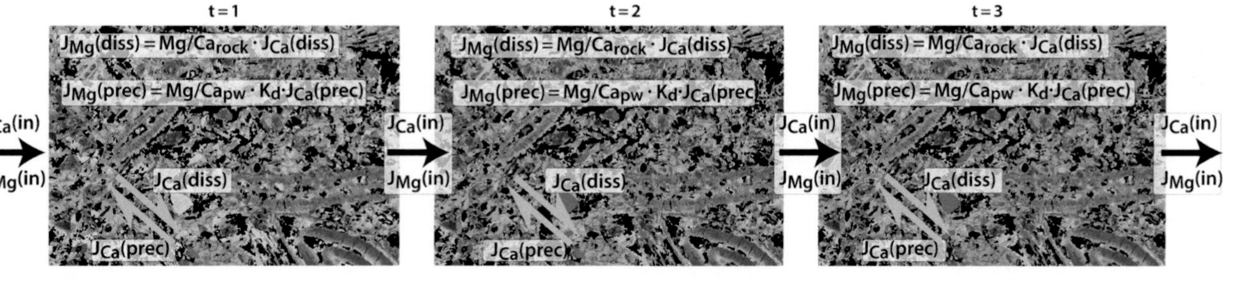

Figure 5 Schematic of a two-box diagenesis model, with the bright colours indicating the progressive recrystallisation at each time step (only three time steps are shown).

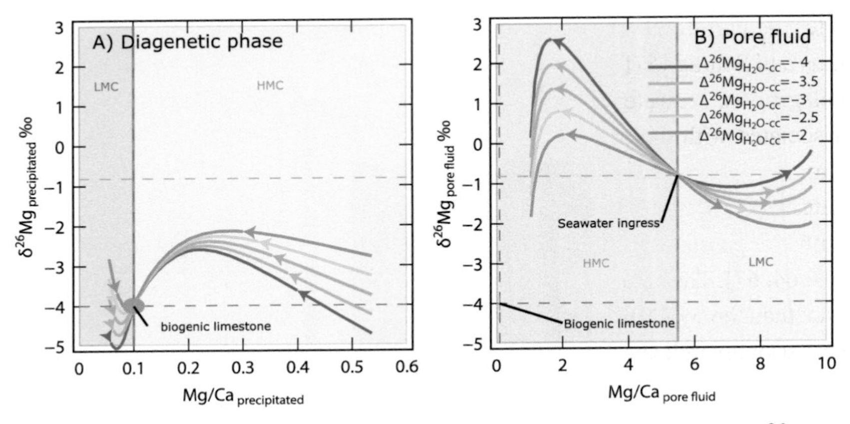

Figure 6 Solutions to a one-box diagenesis model, showing (A) the δ^{26}Mg and Mg/Ca of the total secondary phase that has accumulated, and (B) the pore fluid evolution, from which the secondary phase precipitates. In this case, seawater was assumed to be the fluid which ingresses. The arrows show the direction of evolution with time. LMC is low-Mg calcite and was calculated for a fluid secondary-phase partition coefficient of 0.01, whereas the HMC (high-Mg calcite) scenario was calculated for a fluid secondary-phase partition coefficient of 0.1. The different coloured lines indicate different fractionation factors.

Even for this very simplistic approach to consider diagenesis, there are a wide range of solutions to such a box model depending on the input parameters (Fig. 6). To date, only a handful of studies have applied such models of diagenesis to the Mg isotope proxy [26, 62–64]. The solutions are grouped broadly into four categories: sediment buffered and water buffered, and low-Mg calcite versus dolomitisation. In the case of water-buffered diagenesis, the secondary carbonate minerals which precipitate are controlled by the composition of the advective fluid, which drives recrystallisation. The composition of the secondary phases differs from that of the advective fluid because of a fractionation factor and partition coefficient that control the precise chemistry and isotopic composition of the secondary carbonate. In contrast, with sediment-buffered diagenesis, the composition of the secondary carbonate evolves over time from that of the advecting fluid to that of the dissolving limestone itself, as gradually the composition of the fluid becomes replaced by a supply of Ca and Mg through dissolution (Fig. 6A). The difference between fluid-buffered and sediment-buffered diagenesis depends on the fluid advection rate relative to the dissolution/precipitation rate. The extent to which a carbonate record may be reset by diagenesis therefore depends on a number of factors: the composition of the advecting fluid, the fluid advection rate, relative to the dissolution rate, the partition coefficient and the fractionation factor. For example, Higgins and Schrag [57] argued that diagenesis has only had a small effect on the pelagic ooze seawater record of δ^{26}Mg, mainly because the fractionation factor associated with the recrystallisation of secondary calcite was estimated to be similar to the original fractionation factor between seawater and pelagic carbonate. Other studies have emphasised the complexity of the response of δ^{26}Mg with the possibility of several per mil of variation in δ^{26}Mg being induced [63, 64, 67]. The actual process of dissolution and reprecipitation is more complex than the two-box model discussed here to illustrate the concept [67]. Such scenarios are indicated schematically as solutions to equation 5 (Fig. 6). Indeed, it has now been argued that in some circumstances, δ^{26}Mg values in carbonates may be a tracer of diagenesis rather than providing a record of seawater itself [65, 66].

6 Perspectives for the Future

Despite nearly 20 years of intensive work, the interpretation of the Mg isotope record of seawater is still in its infancy. The modern Mg global biogeochemical cycle is relatively well constrained. Mg isotope ratios have revealed that there must be dolomite removal of Mg [17, 25], though a wide range in absolute values are still possible. Whilst the riverine input to modern seawater is well constrained, the controls on this input are still not fully understood. What is the proportion of silicate- to carbonate- derived Mg? What is the Mg isotope fractionation factor during the formation of clay minerals, and do clays have an affinity for ^{24}Mg or ^{26}Mg [36]? How might this riverine input have changed through time? There is still not a definitive record of seawater δ^{26}Mg through time, nor an accepted archive phase which is free of diagenesis [26, 68]. Two of the three records published so far are in agreement [56–58], perhaps hinting that a definitive seawater record can emerge in the future. The question will then arise as to the extent such a record records seawater through time or is instead a tracer of diagenesis [66]. The next 20 years should be an equally exciting time for the Mg isotope budget of seawater, past and present.

7 Key Papers

7.1 The Mg Isotope Budget of the Modern Ocean

Tipper, E. T., et al., 2006, The Mg isotope budget of the modern ocean: constraints from riverine Mg isotope ratio, *Earth and Planet. Sci. Lett.*, Issues 1-2 241–253. This paper was one of the first to summarise Mg isotope data from the inputs and outputs of seawater. A simple steady-state budget was presented,

but much of the first picture of the oceanic budget is still correct, albeit simplified. One key result is that dolomite must account for at least 8 per cent of the total removal of Mg from the oceans. The most recent update is from:
Shalev, N., et al., 2019, New isotope constraints on the Mg oceanic budget point to cryptic modern dolomite formation, *Nat. Commun.*, 10, 5646. This shows that the low-temperature hydrothermal flux is accompanied by a fractionation of Mg isotopes, implying that the dolomite sink is greater than previously acknowledged, up to 60 per cent.

7.2 The Mg Isotope Budget of the Past Ocean

Pogge von Strandmann, P. A. E. et al., 2014, Modern and Cenozoic records of seawater magnesium from foraminiferal Mg isotopes, *Biogeosciences*, 11, 5155–5168. This was the first published seawater record of Mg isotopes through time. The record is based on foraminiferal calcite, extends back 40 Ma, and, as well as providing a seawater record of δ^{26}Mg, also provides a window into biomineralisation.
Higgins, J. A. and Schrag, D. P., 2015, The Mg isotopic composition of Cenozoic seawater-evidence for a link between Mg-clays, seawater Mg/Ca, and climate, *Earth and Planet. Sci. Lett.* 416, 73–81. The second record of δ^{26}Mg in seawater through time extending back into the Mesozoic. Based on bulk pelagic ooze of different size fractions, the record is very different to that from foraminiferal calcite.
Gothmann, A. M. et al., 2017, A Cenozoic record of seawater Mg isotopes in well-preserved fossil corals, *Geology*, 45, 1039–1042. The third record of Mg isotopes in seawater through time, based on coralline aragonite. The record agrees well with the the bulk pelagic ooze.

7.3 The Role of Diagenesis in Shaping the Geochemistry of the Marine Carbonate Record

There are three key papers that describe approaches to modelling and constraining diagenesis in marine carbonates.
Fantle, M. S. and Higgins, J. 2014, The effects of diagenesis and dolomitization on Ca and Mg isotopes in marine platform carbonates: implications for the geochemical cycles of Ca and Mg, *Geochim. Cosmochim. Act.* 142, 458–481.
Ahm, A.-S. C., et al., 2018, Quantifying early marine diagenesis in shallow-water carbonate sediments, *Geochim. Cosmochim. Act.*, 236, 140–159.
Fantle, M. S., et al., 2021, The role of diagenesis in shaping the geochemistry of the marine carbonate record, *Annu. Rev. Earth Planet. Sci.*, 48, 549–583.

References

[1] E. Goldberg, Chapter 12: Biogeochemistry of Trace Metals, in *Treatise on marine ecology and paleoecology*, ed. by J. Hedgepeth, vol. 1, pp. 345–357 (1957).

[2] S. R. Taylor, S. M. McLennan, *The continental crust. Its evolution and composition* (Blackwell Science, Oxford, 1985).

[3] J. M. Edmond *et al.*, *Earth and Planet. Sci. Lett.* **46**, 1–18 (1979).

[4] L. R. Kump, The Role of Seafloor Hydrothermal Systems in the Evolution of Seawater Composition During the Phanerozoic. In *Magma to Microbe: Modeling Hydrothermal Processes at Ocean Spreading Centers*. Geophysical Monograph Series 178 (American Geophysical Union (AGU), 2008), pp. 275–283.

[5] H. D. Holland, H. Zimmermann, *Int. Geol. Rev.* **42** (2000).

[6] T. T. Isson, N. J. Planavsky, *Nature* **560**, 471–475 (2018).

[7] J. A. D. Dickson, *J. Sediment. Res.* **74**, 355–365 (2004).

[8] A. M. Gothmann *et al.*, *Geochim. Cosmochim. Act.* **160**, 188–208 (2015).

[9] R. M. Coggon *et al.*, *Science* **327**, 1114 (2010).

[10] S. Rausch *et al.*, *Earth and Planet. Sci. Lett.* **362**, 215–224 (2013).

[11] S. T. Brennan *et al.*, *Am. J. Sci.* **313**, 713 (2013).

[12] T. K. Lowenstein *et al.*, *Science* **294**, 1086–1088 (2001).

[13] M. N. Timofeeff *et al.*, *Geochim. Cosmochim. Act.* **70**, 1977–1994 (2006).

[14] J. Horita *et al.*, *Geochim. Cosmochim. Act.* **66**, 3733–3756 (2002).

[15] A. V. Turchyn, D. J. DePaolo, *Ann. Rev. Earth Plan. Sci.* **47**, 197–224 (2019).

[16] H. Elderfield, A. Schultz, *Ann. Rev. Earth Plan. Sci.* **24**, 191–224 (1996).

[17] N. Shalev *et al.*, *Nat. Commun.* **10**, 5646 (2019).

[18] R. A. Berner *et al.*, *Am. J. Sci.* **283**, 641–683 (1983).

[19] L. A. Coogan, S. E. Dosso, *Earth and Planet. Sci. Lett.* **415**, 38–46 (2015).

[20] H. D. Holland, *Am. J. Sci.* **305**, 220–239 (2005).

[21] R. Spencer, L. Hardie, in *Fluid mineral interactions: a tribute to H. P Eugster:* ed. by Spencer, R. J. and Chou, I. M., Geochemical Society Special publication 1990 pp. 409–419.

[22] E. Young, A. Galy, *Rev. Min. Geochem.* **55**, 197–230 (2004).

[23] A. C. Daughtry *et al.*, *Geochim. Cosmochim. Act.* **26**, 857–866 (1962).

[24] A. Galy *et al.*, *Int. J. Mass. Spec.* **208**, 89–98 (2001).

[25] E. T. Tipper *et al.*, *Earth and Planet. Sci. Lett.* **250**, 241–253 (2006).

[26] M. S. Fantle *et al.*, *Annu. Rev. Earth Planet. Sci.* **48**, 549–583 (2020).

[27] E. T. Tipper *et al.*, *Chem. Geol.* **257**, 65–75 (2008).
[28] M. S. Bohlin *et al.*, *Rapid Communications in Mass Spectrometry.* **32**, 93–104 (2018).
[29] C. D. Coath *et al.*, *Chem. Geol.* **451**, 78–89 (2017).
[30] F.-Z. Teng, "Magnesium Isotope Geochemistry". In: *Reviews in Mineralogy and Geochemistry* 82.1 (2017), pp. 219–287.
[31] E. A. Schauble, *Rev. Min. Geochem.* **55**, 65–112 (2004).
[32] E. Young *et al.*, *Geochim. Cosmochim. Act.* **66**, 1095–1104 (2002).
[33] E. A. Schauble, *Geochim. Cosmochim. Act.* **75**, 844–869 (2011).
[34] J. Schott *et al.*, *Chem. Geol.* **445**, 120–134 (2016).
[35] V. Mavromatis *et al.*, *Geochim. Cosmochim. Act.* **114**, 188–203 (2013).
[36] R. S. Hindshaw *et al.*, *Earth and Planet. Sci. Lett.* **531**, 115980 (2020).
[37] W. Li *et al.*, *Earth and Planet. Sci. Lett.* **394**, 82–93 (2014).
[38] F. Wombacher *et al.*, *Geochim. Cosmochim. Act.* **75**, 5797–5818 (2011).
[39] J. A. Schuessler *et al.*, *Chem. Geol.* **497**, 74–87 (2018).
[40] A. G. Dunlea *et al.*, *Nat. Commun.* **8**, 844 (2017).
[41] E. T. Tipper *et al.*, *Global Biogeochem. Cycles* **24**, GB3019 (2010).
[42] E. B. Bolou-Bi *et al.*, *Geochim. Cosmochim. Act.* **87**, 341–355 (2012).
[43] M.Y. H. Li *et al.*, *Earth and Planet. Sci. Lett.* **553** (2021).
[44] S. Opfergelt *et al.*, *Earth and Planet. Sci. Lett.* **341**, 176–185 (2012).
[45] P. A. E. Pogge von Strandmann *et al.*, *Earth and Planet. Sci. Lett.* **276**, 187–197 (2008).
[46] L. Ma *et al.*, *Chem. Geol.* **397**, 37–50 (2015).
[47] T. Gao *et al.*, *Geochim. Cosmochim. Act.* **237**, 205–222 (2018).
[48] K.-J. Huang *et al.*, *Earth and Planet. Sci. Lett.* **359–360**, 73-83 (2012).
[49] S. Opfergelt *et al.*, *Geochim. Cosmochim. Act.* **125**, 110–130 (2014).
[50] J.-S. Ryu *et al.*, *Chem. Geol.* **445**, 135–145 (2016).
[51] J. Wimpenny *et al.*, *Geochim. Cosmochim. Act.* **128**, 178–194 (2014).
[52] K. K. Mayfield *et al.*, *Nat. Commun.* **12**, 148 (2021).
[53] M. J. Mottl, C. G. Wheat, *Geochim. Cosmochim. Act.* **58**, 2225–2237 (1994).
[54] P A. E. Pogge von Strandmann *et al.*, *Front. Earth Sci.* **8**, 109 (2020).
[55] D. P Santiago Ramos *et al.*, *Earth and Planet. Sci. Lett.* **541**, 116290 (2020).
[56] P A. E. Pogge von Strandmann *et al.*, *Biogeosciences* **11**, 5155–5168 (2014).
[57] J. A. Higgins, D. P Schrag, *Earth and Planet. Sci. Lett.* **416**, 73–81 (2015).
[58] A. M. Gothmann *et al.*, *Geology* **45**, 1039–1042 (2017).
[59] P A. E. Pogge von Strandmann, *Geochem. Geophys. Geosyst.* **9** (2008).
[60] C. Saenger, Z. Wang, *Quat. Sci. Rev.* **90**, 1–21 (2014).

[61] W. Broecker, *Am. J. Sci.* **313**, 776–789 (2013).
[62] J. A. Higgins, D. P Schrag, *Earth and Planet. Sci. Lett.* **357–358**, 386–396 (2012).
[63] M. S. Fantle, J. Higgins, *Geochim. Cosmochim. Act.* **142**, 458–481 (2014).
[64] A.-S. C. Ahm *et al.*, *Geochim. Cosmochim. Act.* **236**, 140–159 (2018).
[65] A.-S. C. Ahm *et al.*, **506**, 292–307 (2019).
[66] P F. Hoffman, K. G. Lamothe, *Proc. Nat. Acad. of Sciences.* **116**, 18874–18879 (2019).
[67] R. He *et al.*, *Chem. Geol.* **558**, 119876 (2020).
[68] S. Riechelmann *et al.*, *Geochim. Cosmochim. Act.* **235**, 333–359 (2018).

Acknowledgements

ETT acknowledges funding from NERC grants NE/T007214/1, NE/P011659/1 and NE/M001865/1 and discussions with Will Knapp and Emily Stevenson.

Cambridge Elements

Geochemical Tracers in Earth System Science

About the Series

This innovative series provides authoritative, concise overviews of the many novel isotope and elemental systems that can be used as 'proxies' or 'geochemical tracers' to reconstruct past environments over thousands to millions to billions of years – from the evolving chemistry of the atmosphere and oceans to their cause-and-effect relationships with life.

Covering a wide variety of geochemical tracers, the series reviews each method in terms of the geochemical underpinnings, the promises and pitfalls, and the 'state-of-the-art' and future prospects, providing a dynamic reference resource for graduate students, researchers and scientists in geochemistry, astrobiology, paleontology, paleoceanography and paleoclimatology. The short, timely, broadly accessible papers provide much-needed primers for a wide audience – highlighting the cutting-edge of both new and established proxies as applied to diverse questions about Earth system evolution over wide-ranging time scales.

Cambridge Elements ☰

Geochemical Tracers in Earth System Science

Elements in the Series

Vanadium Isotopes: A Proxy for Ocean Oxygen Variations
Sune G. Nielsen

Cerium Anomalies and Paleoredox
Rosalie Tostevin

Nitrogen Isotopes in Deep Time
Colin Mettam and Aubrey L. Zerkle

The Chromium Isotope System as a Tracer of Ocean and Atmosphere Redox
Kohen W. Bauer, Noah J. Planavsky, Christopher T. Reinhard and Devon B. Cole

The Iron Speciation Paleoredox Proxy
Simon W. Poulton

Reconstructing Precambrian pCO_2 *and* pO_2 *Using Paleosols*
Nathan D. Sheldon, Ria L. Mitchell and Rebecca M. Dzombak

Emerging Patterns in Proterozoic Lipid Biomarker Records: Implications for Marine Biospheric Evolution and the Ecological Rise of Eukaryotes
Gordon D. Love and J. Alex Zumberge

Barium Isotopes: Drivers, Dependencies, and Distributions through Space and Time
Tristan J. Horner and Peter W. Crockford

Lithium Isotopes: A Tracer of Past and Present Silicate Weathering
Philip A. E. Pogge von Strandmann, Mathieu Dellinger and A. Joshua West

Molybdenum as a Paleoredox Proxy: Past, Present, and Future
Stephan R. Hlohowskyj, Anthony Chappaz and Alexander J. Dickson

Iron Formations as Palaeoenvironmental Archives
Kaarel Mänd, Leslie J. Robbins, Noah J. Planavsky, Andrey Bekker and Kurt O. Konhauser

Magnesium Isotopes: Tracer for the Global Biogeochemical Cycle of Magnesium Past and Present or Archive of Alteration?
Edward T. Tipper

A full series listing is available at: www.cambridge.org/EESS

Printed in the United States
by Baker & Taylor Publisher Services